Uncharted, Unexplored, and Unexplained

Scientific Advancements of the 19th Century

Friedrich Miescher
and the Story of Nucleic Acid

Mitchell Lane
PUBLISHERS

P.O. Box 196
Hockessin, Delaware 19707

Uncharted, Unexplored, and Unexplained

Scientific Advancements of the 19th Century

Titles in the Series

Visit us on the web: www.mitchelllane.com
Comments? email us: mitchelllane@mitchelllane.com

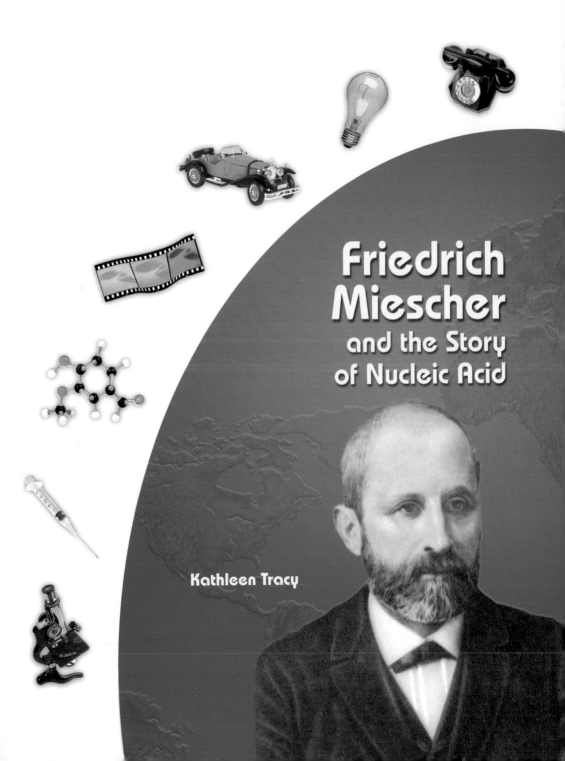

Uncharted, Unexplored, and Unexplained

Scientific Advancements of the 19th Century

Friedrich Miescher
and the Story
of Nucleic Acid

Kathleen Tracy

Scientific Advancements of the 19th Century

Mitchell Lane
PUBLISHERS

Printing   1   2   3   4   5   6   7   8

Library of Congress Cataloging-in-Publication Data
Tracy, Kathleen.
    Friedrich Miescher and the story of nucleic acid / by Kathleen Tracy.
      p. cm. — (Uncharted, unexplored, and unexplained scientific advancements of the 19th century)
    Includes bibliographical references and index.
    ISBN 1-58415-369-5
    1. Nucleic acid—Juvenile literature. 2. Miescher, Friedrich, 1844–1895—Juvenile literature. 3. Physiologists—Switzerland—Biography—Juvenile literature. I. Title. II. Uncharted, unexplored, and unexplained.
QP620.T73 2005
572.8—dc22
                                         2005004252

ABOUT THE AUTHOR: Kathleen Tracy has been a journalist for over twenty years. Her writing has been featured in magazines including *The Toronto Star*'s "Star Week," *A Biography* magazine, *KidScreen* and *TV Times*. She is also the author of numerous biographies including, *The Boy Who Would be King* (Dutton), *Jerry Seinfeld—The Entire Domain* (Carol Publishing), *Don Imus—America's Cowboy* (Carroll), *Mariano Guadalupe Vallejo*, and *William Hewlett: Pioneer of the Computer Age* both for Mitchell Lane. Also for Mitchell Lane, she wrote *Top Secret: The Story of the Manhattan Project* and *Henry Bessemer: Making Steel from Iron*. She recently completed *Diana Rigg: The Biography* for Benbella Books.

PHOTO CREDITS: Cover, pp. 1, 3, 6, 22, 25—Friedrich Miescher Institute; p. 11—NACDL Org.; p. 12—Cancer Research Org.; pp. 14, 16, 17, 36—SCETI: Edgar Fahs Smith Collection; pp. 10, 18, 34—Science Photo Library; pp. 26, 31, 39—Science Research Library; p. 28—Sunergia Medical; p. 37—The Rockefeller University Archives; pp. 40, 41—Sharon Beck.

Uncharted, Unexplored, and Unexplained

Scientific Advancements of the 19th Century

Friedrich Miescher
and the Story of Nucleic Acid

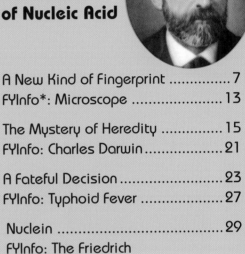

             *For Your Information

Friedrich Meischer was born in Basel, Switzerland. His father, Johann, and his Uncle Wilhelm, were physicians. Although Friedrich originally planned to become a doctor, he later decided to pursue a career in chemistry instead.

Scientific Advancements of the 19th Century

1

A New Kind of Fingerprint

On the cold, clear evening of November 21, 1983, in the English village of Narborough, fifteen year-old Lynda Mann was murdered on her way to visit a friend. Her body was discovered along a desolate path that ran in between a cemetery and a psychiatric hospital. The local police devoted all their resources to solving the crime. But despite following down hundreds of tips, the huge investigation led nowhere. No suspects were ever arrested and the case eventually went cold.

Three years later another fifteen year-old girl, Dawn Ashworth, who lived in nearby Enderby, was reported missing by her family when she never came home from school. The next day, her body was found. Just like Lynda, she had been strangled. And her body was found less than a mile from where Mann had been found.

Within a week, authorities in Enderby had arrested a suspect in Ashworth's murder. A seventeen year-old boy who had been spotted near the scene of the crime was brought in for questioning and confessed to killing Dawn. Police compared evidence samples, and based on blood tests, determined that the same man had in all likelihood killed both girls. So authorities assumed that the boy must have also killed Lynda, even though the suspect denied any involvement.

Just as every person has a unique fingerprint, every person and animal has a unique genetic sequence—or "DNA fingerprint." British molecular biologist Sir Alec Jeffreys (shown here) discovered the method for identifying these sequences. DNA fingerprinting has become an important cornerstone to law enforcement around the world.

To prove he was guilty, the police sent evidence samples to Professor Sir Alec Jeffreys, a DNA expert. Jeffreys had been a medical researcher at Leicester University when he stumbled on a finding that would forever change crime solving and have implications far beyond law.

DNA is made up of long chains of genetic information. In human beings, the vast majority of DNA is the same from person to person. But there are certain sections that are unique to individuals. It's similar to the fact that each person has unique, one-of-a-kind fingerprints. Jeffreys used enzymes as chemical "scissors" to chop up DNA and figured out a way to measure the lengths of the different pieces and show them on an X-ray.

Jeffreys assumed he would find only one or two variable sections. Instead, he was stunned to see something quite different—the DNA from each individual formed a unique pattern. In an article in *The Mail on Sunday*, he described his discovery. "Here was this mass, this welter of unbelievably variable bits of DNA. Looking at that pattern, I suddenly realized that this wasn't just going to pick up the odd bit of slightly variable DNA. This was DNA *fingerprinting*."[1]

It was three years later that he was asked to assist in the Enderby murder. When Jeffreys examined the evidence from the Mann and Ashworth murders he found that the same attacker had indeed, killed both girls. However, the person responsible was *not* the young man who had confessed.

At first Jeffreys thought there was something wrong with the technology. According to *The Mail on Sunday*, Jeffreys said, "But we and the Home Office's Forensic Science Service did additional testing and it was clear . . . he had given a false confession and was released—so the first time DNA profiling was used in criminology, it was to prove innocence."

Even though they had lost their prime suspect, the police now had the real killer's DNA profile. They came up with a unique plan to find the killer. They asked all the men living in the area around the murders to voluntarily give a blood sample so they could be eliminated as suspects. This meant testing over 4,500 people. It was an ingenious plan because if anybody refused, he would immediately become a suspect because anyone innocent had nothing to worry about.

Over a period of eight months, police painstakingly collected blood samples but none of the samples matched the genetic profile Dr. Jeffreys had developed. Finally, police got the breakthrough they had been hoping for. A woman reported that she had overheard a man at work admit he had done a favor for a friend named Colin Pitchfork. Pitchfork had wanted some blood so he wouldn't have to send his own blood to police. Interestingly, police had questioned Pitchfork during their initial investigation because he had a criminal record. But they hadn't found anything suspicious enough to consider him a suspect.

Police immediately picked up Pitchfork again and when his DNA was checked, he matched the samples from the two murders. Pitchfork was tried, found guilty, and sentenced to two life sentences. He was the first person to ever be convicted solely on DNA evidence. "This man would have killed again, no doubt about it," said Professor Jeffreys. "DNA testing helped to save lives."[2]

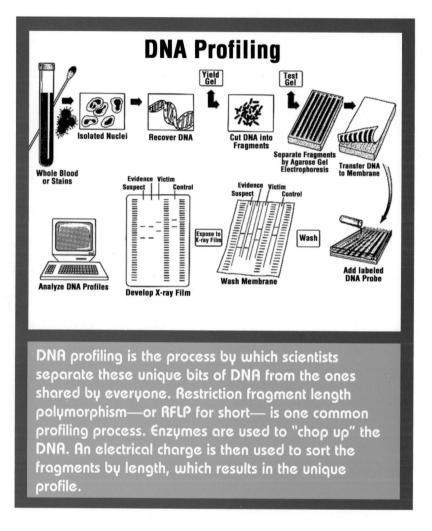

DNA Profiling

Whole Blood or Stains → Isolated Nuclei → Recover DNA → Cut DNA into Fragments (Yield Gel) → Separate Fragments by Agarose Gel Electrophoresis (Test Gel) → Transfer DNA to Membrane

Add labeled DNA Probe → Wash Membrane → Expose to X-ray Film → Develop X-ray Film → Analyze DNA Profiles

Evidence Suspect Victim Control

Wash

DNA profiling is the process by which scientists separate these unique bits of DNA from the ones shared by everyone. Restriction fragment length polymorphism—or RFLP for short— is one common profiling process. Enzymes are used to "chop up" the DNA. An electrical charge is then used to sort the fragments by length, which results in the unique profile.

Within a year of Pitchfork's conviction, DNA profiling was being used by police agencies all over the world. It wouldn't be long before countries would establish national DNA databases which are today major tools in fighting crime. DNA profiling has also been used in other ways, such as in establishing a database to help prosecute those who hunt illegally.

But perhaps one of the most important uses for DNA profiling is to prove the innocence of people wrongly convicted of crimes. According to

Barry Scheck, co-founder of *The Innocence Project*, as of August 2005, 160 people in America have been proven innocent and freed from jail based on DNA testing. Thirteen of those people had been on Death Row awaiting the death penalty.

In his Congressional testimony urging Congress to expand DNA testing, Scheck said, "The pace of post-conviction DNA exonerations has accelerated because states have begun to pass statutes that permit those claiming innocence a chance to gain their freedom. . . . There can be no doubt the number of wrongly convicted freed by DNA testing would dramatically increase if the post-conviction DNA legislation were passed by Congress." Scheck pointed out that not only would innocent people be given back their freedom, but that police could also use DNA to find the real perpetrators. "This is a win-win proposition for law enforcement, innocents who rot in America's prisons and death rows, crime victims, families of all involved, and anyone who loves justice."[3]

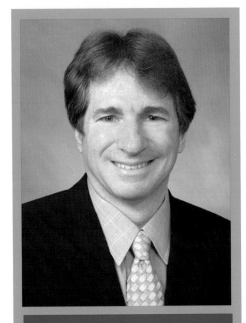

Attorney Barry Scheck co-founded *The Innocence Project* in 1992 with partner Peter Neufeld. They use DNA testing to help people wrongly convicted. As of August 2, 2005, *The Innocence Project* has helped 160 people prove their innocence and be freed from prison.

Since he had helped solve the Enderby murder, Jeffreys kept perfecting DNA profiling. Today, forensics experts only need incredibly small samples to obtain somebody's genetic fingerprint. Thanks to

Professor Jeffreys, Britain developed the world's first and largest national DNA database. It helps solve many cases both new and old cold cases.

But these advances would never have happened if scientists like Jeffreys didn't have an understanding of exactly what DNA is and how it works in the cell. That knowledge began over a hundred years before Jeffreys developed DNA profiling with a discovery by a Swiss scientist who actually didn't realize the importance of his findings. But today, the world recognizes the contribution Friedrich Miescher made in discovering the chemical that became known as DNA.

- crossover
- core
- bifurcation
- ridge ending
- island
- delta
- pore

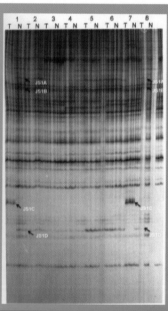

English scientist Sir Francis Galton, along with London police commissioner Sir Edward Henry, developed the first system for identifying fingerprints. Called the Galton-Henry system for fingerprint classification, it was published in 1900. It is still the most widely used system for fingerprint identification in the world. The seven categories of characteristics are listed in the above diagram.

Once the DNA fingerprint has been processed and photographed, scientists can analyze—or type—it. Typing determines the probability of whether another sample is a match. Typing is not just used by police to catch criminals. It can also be used to prove if two people are related.

One of the biggest technological advances that helped scientists unravel the mystery of heredity was the invention of the microscope. Using this valuable tool, scientists were finally able to literally look inside cells.

It had been known since the time of the ancient Romans that looking through certain types of curved crystals made objects appear bigger. It wasn't until the end of the 13th century that this knowledge was used to make the first spectacles to improve the vision of people with bad eyesight.

Sometime around 1590, two spectacle makers in the Netherlands, Zacharias Janssen and his son Hans, accidentally discovered that by using several lenses in a tube they could greatly increase the amount of magnification. What they had created was the first crude compound microscope. Galileo, the famous Italian mathematician, scientist, and inventor who was born in 1564 and died in 1642, learned of the Janssens' discovery and worked out the principles of lenses, including the fact that the more curved the lens, the greater the magnification. This resulted in an improved magnifying tube that had a focusing device.

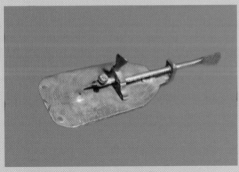

Leeuwenhoek's microscope

Despite Galileo's important discoveries and contributions to microscopy, Anton von Leeuwenhoek is considered the father of the modern microscope. Born in Holland in 1632, von Leeuwenhoek first used magnifying glasses while working in a dry goods store. Hoping to make his job of counting threads easier, he taught himself how to grind and polish lenses. He became so skilled he was able to create a lens so curved it could magnify up to 270 times what the naked eye could see. His discoveries prompted von Leeuwenhoek to quit his job in the dry goods store and work full-time making microscopes and using them to study previously unseen organisms. He was the first scientist to describe microorganisms such as bacteria and blood cells moving through capillaries.

Sadly, none of von Leeuwenhoek's microscopes exist today.

Although Anton von Leeuwenhoek did not invent the microscope, he revolutionized them. His skill at grinding lenses enabled him to produce powerful instruments that magnified over 200 times. He is credited with being the first to identify many microorganisms.

2

The Mystery of Heredity

Since the beginning of civilization, people have known that offspring—whether plant, animal, or human—share characteristics with their parents. The ancient Egyptians, for instance, used this knowledge to develop cross-pollination techniques that allowed them to grow better crops.

Throughout history, there were many ideas as to how exactly traits were passed along. The Greek philosopher, Pythagoras, who lived around 550 B.C., developed a theory that the male's sperm was made up of fluid from all over the body and that the sperm was entirely responsible for his offspring's traits. The mother's contribution was simply to carry the embryo and give birth.

The obvious problem with this idea was that it did not allow for any traits being passed down by the mother. And anyone with eyes could see that children shared traits with *both* parents. So another Greek philosopher, Empedocles, who lived around 460 B.C., suggested that children were actually a product of a blending of the father and mother's genetic material.

One of the best-known of the Greek philosophers was Aristotle, who is believed to have been born in 384 B.C. and who died in 322 B.C.

Aristotle came up with a theory that would be accepted for almost two thousand years. According to Aristotle, sperm was purified blood. He also believed a man's blood was more pure than a woman's. So the combination of sperm and a woman's menstrual blood is what created the embryo. Because, he said, a man's blood was more pure, sperm was the actual source of life while the woman's blood contributed to the actual formation of the child.

The Greek philosopher Aristotle is considered one of the greatest thinkers of all time. He developed a theory that menstrual blood and sperm combined to create an embryo. He believed that sperm was the source of life. Even though his theory was ultimately proven wrong, it was accepted as fact for nearly two thousand years.

The trouble with all of these theories is that there was no way to test them to see if they were true. Thanks to the development of the microscope by Anton von Leeuwenhoek, scientists were finally able to see inside cells, unravel the mystery of how traits are passed from parent to child, and uncover the mechanism by which all life evolves. Research by Leeuwenhoek and William Harvey in the 17th century led to the discovery that female animals produced eggs. From there it was shown that it was the union of the female's egg and the male's sperm that resulted in an embryo.

But there was still disagreement as to what the egg and the sperm contributed. Some believed that a whole human being was contained in each sperm and that the mother was, again, just there to provide food and a way to be born. Others took the opposite view and believed the woman's eggs each contained an embryo and the sperm simply prompted the egg to start developing. Eventually, some scientists began to doubt that human beings were "preformed" as suggested by these theories. Kaspar Wolff proposed that the egg and sperm actually contained "particles" of some kind that directed how the embryo would grow.

In November 1859, Charles Darwin published *On the Origin of Species*. In it, he described his theory of evolution. According to Darwin, all species evolved, or changed, over the course of time. What caused species to change was something Darwin called natural selection, which meant that species would develop traits over a long period of time that would enable them to best survive whatever their environment was. Darwin's theory was brilliant in its simplicity. It also explained why there was such diversity, such as why a type of snake that lived in the desert

British naturalist Charles Darwin developed the theory of evolution while on a five-year science expedition aboard the *HMS Beagle*. His observations of wildlife in the Galapagos Islands led him to conclude species changed over time because of natural selection. These theories explained the vast diversity seen within the animal and plant kingdoms.

had traits a snake that lived in a damp forest didn't have. But the one thing Darwin couldn't explain was the mechanism that caused this to happen.

The proof Darwin needed to support his theory would be found in a patch of pea plants. Gregor Mendel was born in Austria in 1822 and

Gregor Mendel's work with pea plants proved individual traits were passed down from parents to offspring equally. But his theories of heredity were so contradictory to the conventional wisdom of the time they were initially rejected by other scientists. It would take over thirty years before his work was accepted and appreciated for its brilliance.

grew up on his family's farm. Because they didn't have much money, his parents couldn't afford to send Gregor to a university. To get his education, Gregor entered a monastery. Four years later he took his vows and became an Augustine monk. He went to live at a monastery in Brno, located in what is now Czechoslovakia.

All his life Gregor had loved learning and studied many different subjects, including meteorology (the science of weather), but his two favorite subjects were different theories of heredity and plants. One day while he was strolling around the grounds of the monastery he got an idea that would combine his favorite areas of study—plant life and heredity. He decided to conduct experiments to figure out why pea plants bloomed the way they did.

In his first experiments, he grew two different types of plants next to each other to see if just being near each other would cause them to exchange traits. He quickly realized that they wouldn't. No matter how close they were or what other kinds of plants were nearby, every generation of plants retained its distinct traits, completely unaffected by its environment. To Mendel, that meant that whatever made plants change must be located inside of them.

Mendel didn't necessarily set out to prove that idea by breeding pea plants. Initially, he just bred the plants because he thought it was fun. But over the course of several plant generations, he noticed that there seemed to be a mathematical ratio to certain traits. Some traits seemed to "dominate" others, such as color or leaf variety.

Based on these observations, Mendel formulated a theory as to why this happened. To prove he was correct, between 1856 and 1863 he crossbred thousands of pea plants and kept detailed records of his results. In the beginning, he focused on one trait at a time. But it seemed that no matter what trait he was studying, the results of his breeding always yielded one of three ratios: 100 percent of the offspring were identical to one "parent" plant; 75 percent were identical to one parent or 50 percent were like one parent.

Using pea plants, Mendel was able to establish precise mathematical ratios for traits passed down. Based on these observations, he realized that some traits were stronger, or dominant, and some traits weaker, or recessive. He also determined that traits don't blend together but are passed down intact.

Some traits seemed to overpower other traits all the time. For example, when he crossed a tall plant with a tall plant, the ratio was usually 75 percent tall and 25 percent short in the offspring. When crossing a tall plant with a short plant, the ratio was 50 percent

—50 percent. And crossing a short plant with a short plant *always* resulted in a short plant.

These ratios remained remarkably consistent and led Mendel to conclude that the outward appearance, or phenotype, of an offspring was determined by its genotype, or genetic factors. Mendel also theorized that hereditary factors do not combine, but are passed intact. Specifically, he believed his observations proved that each parent transmits only half of its hereditary factors to each offspring. The reason he got the ratios he did, according to Mendel, was because some traits are dominant, or stronger, while others are recessive, or weaker.

In 1866, the year after the American Civil War ended, Mendel published *Experiments with Plant Hybrids*, in which he described how traits were inherited: both parents contribute material that determines the characteristics of their offspring. Despite his meticulous records and logical conclusions, Mendel's theories were shockingly ignored and quickly forgotten by the scientific community at large. As has happened so many times in science, his ideas were so brilliant and unprecedented that people simply couldn't—or wouldn't—believe him. Because Mendel became the abbot, or leader, of his monastery in 1868, he no longer had time to conduct research and it would take thirty-four years before his work was rediscovered.

But eventually, Mendel would be recognized as the "father of genetics" and his theories would tie together an explosion of discoveries that turned biology into a respected science.

According to Dr. Martin Hewlett, "In retrospect, the sixth decade of the nineteenth century was truly remarkable with respect to the development of the science of biology. By the end of those ten years . . . the key elements of what would become modern biology had been discovered and formulated."[1]

But it wouldn't be until a Swiss chemist named Johann Friedrich Miescher isolated an unknown substance from discarded bandages that the true future of modern biology would begin.

Since the dawn of civilization, people had tried to understand how traits are passed from one generation to the next and why the same type of animal developed different traits depending upon where that type lived. It wasn't until Charles Darwin came up with the then-radical theory of evolution by natural selection that the role of heredity took center stage.

Born in England on February 12, 1809, Charles Robert Darwin had originally set out to be a doctor. But he soon discovered that the sight of blood made him sick. Instead, he decided to be a minister. But while studying religion at Cambridge University, he developed a fascination with nature. When he was 22, he joined an expedition that was planning to sail around the world, and on December 27, 1831, Darwin boarded the HMS Beagle on the trip that would change the world.

The official purpose of the expedition was to map both coasts of South America, but Darwin used the voyage as a way to study natural life in different parts of the world. At the time, most people in Europe believed that the earth and all the animals on it had been created by God as described in the book of Genesis in the Bible. That meant all living creatures were exactly the same as they had been when created. But Darwin had different ideas. For one thing, fossil records indicated that many of the animals alive today hadn't existed millions of years ago. As he traveled and saw the extraordinary diversity of plant and animal life on the planet, he began to develop his own theory for the origin of life on earth.

The defining moment for Darwin came when they arrived at the remote Galapagos Islands off the western coast of South America. There he saw a unique collection of animals and plants, including many species not found anywhere else in the world. His observations led him to theorize that animals adapt to their surroundings and that the animals that adapt the best, survive. The traits that allow them to survive were passed to their offspring. Now we know those traits are passed via DNA, but at the time, Darwin didn't know the mechanism of heredity, just its outcome.

The implication of this was staggering. It meant that all life on earth was the product of ongoing evolution. Not surprisingly, Darwin's theory was controversial because it defied the Bible's version of creation. Some people were also upset by the notion that man was suddenly no more than an evolved ape. Darwin died on April 19, 1882. But as the mysteries of DNA were uncovered, Darwin's theories were validated and would prove to be a stunning example of how simple observation can be the most important scientific tool.

Friedrich Miescher (shown here with his parents) was born into a well-respected family of prominent physicians. Unlike his outgoing father, Friedrich was shy as a boy and preferred to be alone. He would spend his time reading or just sitting and thinking.

3

A Fateful Decision

The canton, or state, of Basel lies in Northern Switzerland. Bounded on the north by the Rhine River and on the south by the Jura Mountains, it is an area of great natural beauty, filled with fertile fields, meadows, forests, and orchards. Most of the residents speak German. In 1833 the canton was split into two independent half-cantons. Basel-Land and Basel-Stadt, the latter made up mostly of the capital city Basel and its surrounding suburbs. Basel is a major Swiss financial center and is both an important rail hub and port city. It is also home to the nation's chemical and pharmaceutical industries. The University of Basel was founded in 1460 by Pope Pius and attracted some of the brightest minds in Europe, earning the city a reputation of an intellectual center. It was here that on August 13, 1844, Johann Friedrich Miescher was born.

Johann Miescher's father Friedrich was a doctor and taught pathological anatomy at the University. His uncle, Wilhelm His, was a famous embryologist so Friedrich's family was well respected and considered important members of the city's intellectual elite. The young Johann Friedrich, who was called Fritz by his family and friends, was a shy boy who enjoyed spending time alone just thinking. Part of his shyness may have been due to the fact that he suffered from impaired hearing. But he was very intelligent and was an excellent student. Fritz's father was a talented singer and often gave public concerts, and like his father, Fritz had a deep love and enjoyment of music.

Since Fritz had been born into a family of doctors, it was assumed that he would also go into the profession. In the 19th century, the field of medicine was very different from modern medicine. Scientists were only beginning to understand the true inner workings of the human body, so it was an exciting time because so many discoveries were happening, seemingly on an almost weekly basis. But Fritz initially was cool to the idea of being a doctor. Instead, he wanted to be a priest. This idea did not sit well with his father, so Fritz gave in to his father's wishes and agreed to pursue a career in medicine.

During the summer of 1865 while he was still a medical student, Fritz spent time in the city of Göttingen so he could work in the laboratory of Adolf Strecker, who was an organic chemist. His time there passed uneventfully but when he returned to Basel, Miescher contracted typhoid fever. He became so ill and weak that he had to drop out of school. It took him almost a year to recover. Eventually, he was strong enough to return to school and finally earned his degree in 1868 when he was twenty-four years old.

But while his fellow graduates celebrated by singing songs and drinking wine, Fritz watched the festivities with a heavy heart. He now needed to figure out what specialty in medicine he should pursue. So he wrote a long, detailed letter to his father explaining his thoughts and concerns about the future. He carefully wrote out all his options followed by both the plusses and minuses of each. What was troubling Fritz the most was that he felt that his hearing impairment would prove to be too great a handicap to overcome if he tried to work in a medical field that required him to deal with patients.

According to the book *Girders in the Sand*, his uncle Wilhelm wrote him a letter encouraging Fritz to follow his heart.

Dear Fritz,

I cannot doubt your aptitude for work, and I'm certain your achievement will be high, if your optimistic mood is well maintained. Your own self confidence and dedication are all you require; not empty, foolish, over-confidence that cannot make mistakes—but to aspire to self-reliant, continuous work: giving of your best, you will become the best. Though doubts in other fields will still lurk, your individuality from the rest will be assured; so take from this good heart and choose your field— then set yourself apart.[1]

In the end, Fritz decided that instead of being a general practitioner or surgeon, he would go into chemical research. Heeding his uncle's advice to be the best he could be Fritz decided to study with the top men in the field. In 1868, Fritz went to work in a newly established lab, located in the castle at Tübingen, that was run by Ernst Hoppe-Seyler. A well-known German physician, Hoppe-Seyler was a pioneer in the field of biochemistry and was famous for a number of important discoveries.

He was the first to isolate lecithin in its pure form. Lecithin is a lipid, a kind of natural fat, produced by the liver and is one of the key building

After receiving his medical degree, Miescher decided to pursue a research career. He accepted a job in a lab run by Felix Hoppe-Seyler. A pioneer of modern biochemistry and molecular biology, Hoppe-Seyler inspired the young scientists working with him and many important discoveries were made in this lab (shown here), which had been built inside a castle.

blocks of cell membranes; without it, they would harden. Lecithin is also the primary compound that makes up the protective sheaths surrounding the brain and helps keep the circulatory system healthy.

Hoppe-Seyler is also credited for coining the term "proteid," which we now call protein. But his most notable achievement was in figuring out how cells in the body "breathe."

In the early 19th century scientists believed that oxygen was distributed from the lungs to body tissue via the water in the blood. By the 1850s, most researchers had concluded that the oxygen must move throughout the body in combination with some other as-of-yet unknown substance. It was Hoppe-Seyler who finally discovered the actual mechanism of how oxygen traveled inside the human body. He identified hemoglobin, which gives red blood corpuscles its color, as the substance that oxygen attaches to. Once the hemoglobin reached the tissue, it diffused out of the blood into the cells where it combined with carbon and hydrogen to form carbon dioxide and water. The energy given off during this process was literally the energy needed to sustain life.

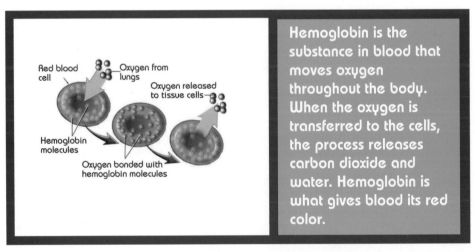

Red blood cell

Oxygen from lungs

Oxygen released to tissue cells

Hemoglobin molecules

Oxygen bonded with hemoglobin molecules

Hemoglobin is the substance in blood that moves oxygen throughout the body. When the oxygen is transferred to the cells, the process releases carbon dioxide and water. Hemoglobin is what gives blood its red color.

Fritz was inspired to be working with such an esteemed scientist, and immediately set out to design the experiments he wanted to conduct. Little did Fritz know he was about to stumble on one of the most important discoveries in history.

During the 19th century, diseases such as typhoid fever, which Fritz contracted as a young man, killed thousands of people every year. During the United States' Civil War, 1861–1865, it is estimated almost as many soldiers died from typhoid and dysentery as were killed in battle.

Typhoid Fever is spread by the Salmonella Typhi *bacteria*. This kind of germ lives only in human urine and feces. Because the bacteria resides in a person's intestinal tract it can be contracted, or caught, by eating food or drinking water contaminated with infected sewage. People can also directly spread the disease to others by not washing their hands after going to the bathroom.

Salmonella Typhi

When Fritz contracted typhoid, nobody knew what caused it, nor were there any antibiotics to treat it as there are today. So anyone who came down with typhoid had to suffer as the illness ran its course. A person usually began to show symptoms anywhere from one to three weeks after being exposed to the bacteria. The first signs of the disease are a high fever, terrible headache, weakness, fatigue, and a sore throat.

During the second week, many patients developed a red rash for a few days, one of the telltale signs that they had typhoid and not something else like the flu. It is during the second week that Fritz would have become extremely ill. His fever would have remained high and he would have suffered from abdominal pain. Many patients develop diarrhea and lose a lot of weight, making them even more physically weak.

After the third week patients go into what is called the typhoid state, which is when the patient is basically motionless because they are so weak and exhausted. It is during this phase that most people died. But those who survive past this stage of the illness usually begin to recover.

A month into the illness, the fever finally begins to subside, or decrease, and within two weeks, is back to normal. However, even though the disease itself is over, it would usually take people a long time to completely regain their strength and health.

It is interesting to note that the scientist Fritz went to work with, Adolf Strecker, participated in developing a treatment for another deadly 19th century disease, malaria. It was Strecker who determined the proper chemical formula for quinine, which was the only effective treatment for malaria until the 1930s.

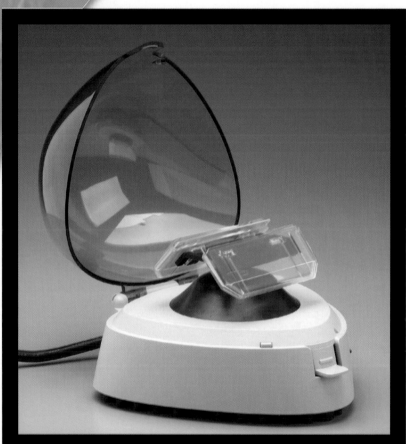

A centrifuge is a piece of laboratory equipment. It is used to separate solids suspended in a liquid sample. It works by spinning the samples very fast so that the heavier substances are forced to the bottom of the liquid.

4

Nuclein

Fritz had decided he wanted to research the chemistry of the cell nucleus, which is known as histology, or the study of the microscopic structure of cells. He had been greatly influenced by his uncle, who once said, "Since in my own histological research I had come to the conclusion again and again that the ultimate questions about the development of tissues can be solved only by way of chemistry."[1]

This was the specialty of Hoppe-Seyler's lab. Separating out a cell nucleus, the sphere-shaped mass within a cell, enclosed by a membrane, was typically difficult because of its small size. But white blood cells are different because they have unusually large nuclei (plural of nucleus) and not much cytoplasm, which is the clear, jelly-like substance found inside cells. Fritz initially had wanted to study the composition of lymphocytes, a type of white blood cell. But extracting these cells from a person's lymph glands is not easy and even when it works, a scientist couldn't get very many to work with. So instead, Fritz decided to study leukocytes, another type of white blood cell that is present in large quantities in the pus found in infections.

So Fritz went to a nearby hospital that was caring for many soldiers recovering from wounds sustained in the on-going Crimean War, in which England and France were fighting to prevent Russia from claiming territory in Europe. The war lasted from 1854 to 1856. All over the hospital were discarded bandages filled with blood and pus. So every day he collected an armful of bandages and took them back to the lab.

Now that Fritz had his source for leukocytes, his next challenge was to wash the cells without damaging or destroying them. First, Fritz used a number of different salt solutions. While the cells were cleaned, they became too swollen. After much trial and error, Fritz came up with a solution made of diluted sodium sulfate that allowed him to successfully rinse well-preserved cells off of the bandages.

The next step was to filter that solution in order to remove tissue fibers from the cells. Then he left the beaker holding the filtered solution, and the cells eventually settled at the bottom. Today, scientists use a centrifuge, a machine that spins at high speed, to separate components in a matter of seconds. But centrifuges hadn't yet been invented in Fritz's day so he had to wait for the cells to settle naturally. However, to Fritz it was worth the wait. When he examined them under a microscope, the leukocytes seemed intact and showed no sign of damage.

Next, he needed to separate the nuclei from the cytoplasm, something that had never been done before. It took him long hours of hard work. Fritz tried several methods. He finally treated the cells with warm alcohol to remove lipids, and then removed the cytoplasm's proteins with an enzyme, pepsin. A gray precipitate, or solid substance, was left. Fritz then purified the precipitate in the same way he had the original pus cells, and the result was another precipitate. It was obvious this material—whatever it was—had come from the nuclei of the leukocytes. So Fritz called it "nuclein."

Through a series of experiments, he determined that the nuclein was made up of nitrogen, phosphorus, and sulfur. He also determined that its molecules were very large. Because it was so relatively rich in phosphorus and because pepsin didn't eat away at it, Fritz ruled out the possibility it was a protein. Proteins usually don't contain much phosphorus. At the time that Fritz was working on his cells, three

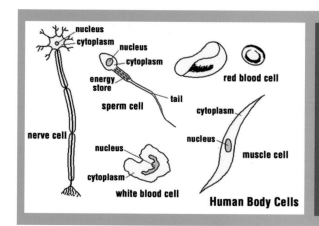

nucleus
cytoplasm

nucleus
cytoplasm

energy store

red blood cell

sperm cell

tail

nerve cell

cytoplasm

nucleus

nucleus

muscle cell

cytoplasm

white blood cell

Human Body Cells

There are many different types of cells in the human body, as listed in the diagram. Each type of cell has a different, specialized function. But all cells carry an individual's DNA code within the nucleus.

categories of substances had been identified in the human body. These were lipids, or fats, polysaccharides, which are sugars and starches, and proteins. So the fact that Fritz had discovered another substance in the cell's nucleus was a dramatic and monumental finding—so much so that Hoppe-Seyler didn't actually believe it at first. He would not let Fritz publish or reveal his findings until Hoppe-Seyler himself repeated the experiments and came up with the same results. Finally, Fritz was able to publish his discovery although nobody really understood yet the importance of his work. His paper—like Mendel's—was almost immediately forgotten.

In 1870, Fritz returned to his hometown of Basel and was named a professor of physiology at the University—the same position that both his uncle and father had held. The upside was that the job would give Fritz more money for research and an equipped lab to work in; the downside was that he had to teach. Unfortunately, he was not a very good teacher. Besides still being so shy, he was so preoccupied with his work that he simply didn't devote much of himself to his students.

But his research continued to show results. The Rhine River at the time was famous for its salmon as a food source, but for Fritz, the salmon was important for another reason. While all sperm cells have large nuclei, salmon sperm is especially big, making it excellent material for Fritz's work.

The "nuclein" Fritz had extracted from the leukocytes was actually a blend of DNA and RNA. But working with the larger sperm nuclei, Fritz and the students in his laboratory were able to extract pure DNA in 1874 when they separated "nuclein" into a protein and acid. Now able to ascertain its true composition, one of Fritz's pupils, Richard Altmann, coined the name "nucleic acid."

It didn't take long for Fritz and other scientists to determine that nucleic acid was present in all cells. It became the fourth category of substances found in the human body, along with fats, carbohydrates, and proteins. But the actual function of nucleic acid remained a mystery. However, the majority of scientists believed that the nucleus of a cell was responsible for holding the key to heredity even if they didn't know how.

Fritz himself had a suspicion that nucleic acid was involved in the process. In a letter written to his uncle in late 1892, Fritz observed that he had noticed that some of the large molecules seen during his research were made up of chemical pieces that were similar although not identical. He observed that it was, "Just as the words and concepts of all languages can find expression in twenty-four to thirty letters of an alphabet,"[2] meaning that perhaps heredity could be determined by varying a relatively small number of compounds. In essence, what he was suggesting was a genetic code.

Although he had gained notoriety for discovering nucleic acid, the rest of Fritz's career attracted little attention because his research did not result in any new information on nucleic acid. But he remained dedicated to his profession, and in 1885 Fritz founded Switzerland's first physiological institute, which continues to this day.

For the rest of his professional life Fritz continued to study nucleic acid on and off. While doing some experiments regarding the chemistry of fertilization—which was still a great unknown—Fritz reportedly suggested that nucleic acid very probably played a part. But for unknown reasons, he never pursued this theory.

But other scientists did. And what they would discover was the answer to the puzzle of heredity.

Institute

Even before the discovery of nucleic acid, the city of Basel, Switzerland, had been known as a leading learning center, mostly because of its university. But in the years since Friedrich Miescher made his important findings, Basel has become home to several important scientific institutions that attract researchers the world over, as well as to numerous pharmaceutical companies involved in the development and manufacture of new medicines.

The most famous of these scientific institutions is the Friedrich Miescher Institute. It was founded in 1970 with the mandate, or purpose, to sponsor and promote advances in biochemistry and medicine by providing a thriving research environment. The current main areas of research at the institute are growth control, neurobiology, and epigenetics, which is the study of the changes that occur in genes during a cell's development and when it replicates, or makes copies of itself. This research is important in potential treatments for diseases such as cancer, Alzheimer's Disease and diabetes.

In addition to its research, the Friedrich Miescher Institute also offers educational programs for students, including a Ph.D. program that is run in conjunction with the University of Basel. The goal is to give students as much practical and real life experience doing research.

Basel is also home to the Swiss Tropical Institute. Founded in 1943, the STI is another teaching and research center. Its focus is the study, understanding, prevention, and treatment of infectious diseases, which includes everything from malaria to Dengue Fever. Like the Miescher Institute, the STI is affiliated with the University of Basel and many of the researchers also teach at the school.

However, one of Basel's most famous medical discoveries was made quite by accident. In April 1943, Dr. Albert Hofmann was working in Sandoz, one of Basel's top pharmaceutical companies. Hofmann was working with a type of fungus in hopes of discovering a cure for migraine headaches when he began to feel strange. Thinking he was sick, he went home but when he lay down and closed his eyes, he "saw" colorful, fantastic images in his mind's eye. He didn't know it, but Dr. Hofmann had just taken the first known "acid trip" because he had inadvertently created lysergic acid diethylamide, or LSD, from the fungus, and had unknowingly absorbed the drug through his fingertips.

For years after, scientists would study the mind-altering effects of the drug. But it was a dangerously powerful one, and when it became popular among some young people during the mid-1960s, Congress passed a bill in 1966 making LSD an illegal substance in the U.S. Today there are still some scientists who believe that there could be genuine psychiatric medical uses for LSD, but because the drug is now universally outlawed, no research can be openly conducted.

In 1953 American James Watson and British-born Francis Crick solved the mystery of DNA while working together at Cambridge University. They determined that DNA was a brilliantly simple double helix that replicated by "unzipping" itself down the middle. Their work earned them the Nobel Prize in 1962 and lifelong fame.

5

Miescher's Legacy

Fritz had recognized that nucleic acid combined in the chromosomes with protein and other materials. Because protein is so essential to every life process, he and many other scientists assumed it would eventually be proven that proteins somehow carried the hereditary information of an organism. Part of that belief was based on the knowledge that individual proteins are made up of a combination of twenty amino acids—a kind of chemical "alphabet" which theoretically could be combined into almost infinite combinations—that could carry our genetic make-up. But they eventually found out they were wrong. In 1893, Albrecht Kossel learned that nucleic acid was actually made up of four chemical compounds, adenine (A), thymine (T), guanine (G), and cytosine (C)—all of which contained nitrogen.

But Fritz would not live to see where that discovery would lead. Two years later, in 1895, Fritz's health began to fade. Although he had always been a workaholic, he suddenly barely had the strength to get through the day. When he was diagnosed with tuberculosis, Fritz was forced to retire. He died later that year in the Alpine city of Davos, which has since become a popular ski resort. It is interesting to note that while he was alive, Fritz never published much of his work. It was only after his death that colleagues and friends gathered his work and had it published or much of it might have been lost forever.

German physician Albrecht Kossel was the first to identify that nucleic acid was made up of four distinct compounds: adenine, thymine, guanine, and cytosine. For identifying the nucleic substances, Kossel won the 1910 Nobel Prize in Medicine.

Because it also consists of the sugar deoxyribose, Fritz's nuclein became known officially as deoxyribonucleic acid, or DNA. And as more scientists devoted their energies into unraveling the mystery of DNA and heredity, the discoveries kept adding pieces to the puzzle. In the 1920s, an English bacteriologist named Fred Griffith discovered a curious occurrence while working with pneumonia bacteria. He had two types of bacteria—a mutant form that could not cause illness and another virulent, or disease causing, type that could. In the experiment, he killed the virulent bacteria by boiling it. Once dead, these bacteria were no longer capable of causing pneumonia. However, when Griffith mixed the dead bacteria with the mutant bacteria, the mutant bacteria suddenly became virulent. Somehow, the dead bacteria had chemically passed along the ability to cause disease to the formerly harmless bacteria. Griffith called this the "transforming principle." Today we know this chemical unit as a gene.

Twenty years later, in the 1940s, Oswald Avery and his team followed up on Griffith's discovery. Working at the Rockefeller Institute, Avery found that a pure extract of the "transforming principle" was not affected at all by protein-digesting enzymes. However, it *was* destroyed by a nucleic acid digesting enzyme. What this proved was that the "transforming principle" was actually made up of DNA. The inference of

Canadian scientist Oswald Avery was the first to identify DNA as the genetic material. At the time, most scientists believed that proteins were responsible for heredity. But soon after publishing his findings, other scientists were able to confirm that DNA is how traits are passed on from parent to offspring.

this to Avery was clear: DNA, not protein, was the genetic carrier. Although not everyone immediately accepted his conclusions, it was only a matter of time before his findings were confirmed.

Scientists soon discovered that different species have different amounts of the found DNA bases—A,G,C,T—and that the ratio of A to T and G to C was *always* the same. In other words, every molecule of adenine always had a corresponding molecule of thymine in DNA, and every molecule of guanine had a corresponding cytosine. This seemed to strongly suggest that these "base pairs" were somehow importantly connected. What exactly that connection was would be revealed a decade later.

Thanks to the development of X-ray techniques, it was now possible to actually look at DNA, which suggested DNA had a kind of corkscrew shape. In 1953, James Watson and Francis Crick built several experimental models trying to figure out exactly how DNA was structured. What they eventually deduced was breathtaking in its simplicity of design and stunning in its power to create all of life's diversity.

What Watson and Crick realized was that DNA must be constructed as a double helix, or a kind of twisted ladder shape. The sides were a "spine" made up of the sugar and phosphate molecule and the "rungs" of the ladder were the base pairs. That's how C could always combine with G, and A to T. This structure also explained how DNA replicated, or reproduced, itself. When the helix "unzipped" down the middle, two identical double helixes were then formed because bases would join with its complementary partner to form a new helix.

This discovery "was a complete watershed—the great change of the last century in biology," said Nobel-prize-winner Sidney Brenner of the Salk Institute during an interview with *The Cincinnati Post.* "I know that when I saw their model for the first time it was like turning over a page. It was, *Well, you can forget about everything else. Let's get on with this.*"[1]

Once the structure of DNA was understood, it opened the door for many other discoveries, such as our understanding of how RNA, or

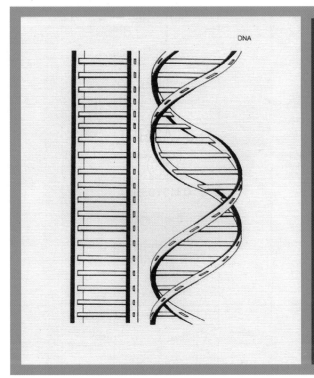

DNA

Watson & Crick used large 3-D models to figure out the structure of DNA. Although it was also known that the four bases always combined in equal ratios of Adenine(A) and Thymine (T) and equal rations of Guanine (G) and Cytosine (C). The double helix, which is the shape of a twisting ladder, showed how the base pairs made up the "rungs," which was how C could always combine with G, and A could always combine with T.

ribonucleic acid, assists in the manufacture of our body's proteins by carrying the encoding information from DNA to the cells. Today, scientists are working on the Human Genome Project, in which they will catalogue the approximate 35,000 genes humans carry on their 26 chromosomes. All of this is a direct result of Fritz's 1869 discovery that for so many years went almost unnoticed because nobody realized its importance. Speaking during a Nobel Prize ceremony in 1968, Professor P. Reichard noted, "During the 19th century the Nobel Prize had not been established. Had the prize existed it is unlikely that it would have been awarded for the discoveries of nucleic acids and genes."[2]

However, today the world knows just how important Fritz's contribution was. So in 1974 Switzerland opened the Friedrich Miescher Institute in Basel, which is dedicated to biomedical research, to honor the man whose discovery was the first step in unraveling the mystery of the code of life.

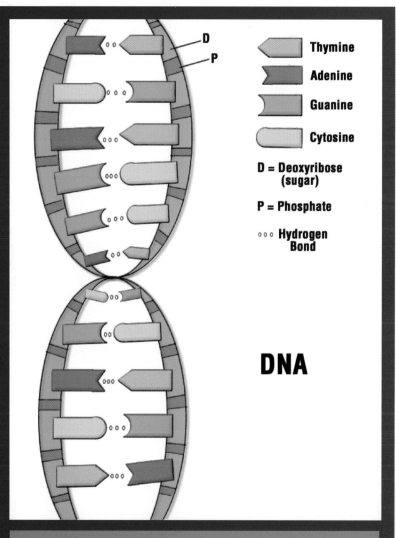

Thymine	
Adenine	
Guanine	
Cytosine	

D = Deoxyribose (sugar)

P = Phosphate

○○○ Hydrogen Bond

DNA

DNA is made up of four base compounds: Adenine, Thymine, Guanine and Cytosine. These four bases always bind together in A-T and C-G pairs, held together by molecules of hydrogen. He outer edges, or spine, of the DNA double helix, are made up of sugars and phosphate.

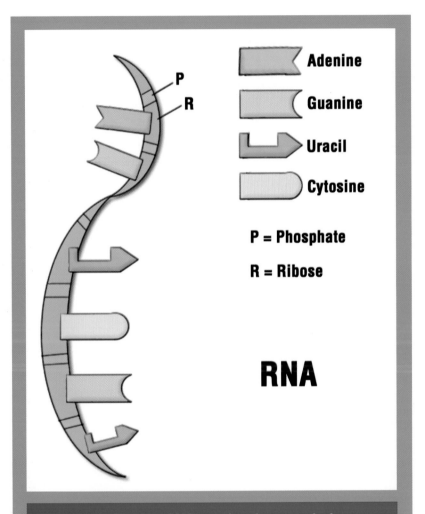

Adenine

Guanine

Uracil

Cytosine

P = Phosphate

R = Ribose

RNA

Ribonucleic Acid, or RNA, carries the encoded information stored in DNA to the cells. Specifically, it assists in the manufacture of proteins. Although similar in chemical make-up to DNA, RNA is distinguished by having molecules of Uracil instead of Thymine.

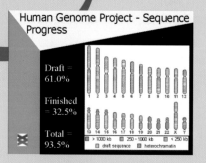

Human Genome Project - Sequence Progress

Draft = 61.0%

Finished = 32.5%

Total = 93.5%

> 1000 kb   250 - 1000 kb   < 250 kb
draft sequence   heterochromatin

In 1990, researchers from all over the world started an ambitious effort. The goal was to map and sequence the genomes, or entire collection of genes, for a number of living organisms including yeast, the nematode worm, mice, and humans. Called the Human Genome Project (HGP), it was made up of scientific teams from six countries—the United Kingdom, France, Japan, China, Germany and the United States—and was organized by the United States Department of Energy and the National Institutes of Health. The purpose of sequencing our DNA was to better understand the specific functions controlled by different parts of the DNA code.

When the project was first announced, it was estimated that it would take at least fifteen years to finish the project. However, nobody anticipated that technology would develop as quickly as it did. So after just thirteen years, researchers were able to announce they had successfully identified the sequence of the more than three billion base pairs that comprise the approximately 25,000 genes in human DNA. Ironically, 2003 also happened to be the 50th anniversary of Watson and Crick discovering DNA's double helix structure.

Thanks to the HGP, scientists were able to decode the genetic instructions that guide our development from the moment we are conceived to when we age and die. Prior to the project, scientists thought each gene performed a single function. But now they realize that individual genes have the capability to perform more than one function, meaning our genetic make-up is even more complex that originally anticipated. Since it is our genes that make us more prone to develop or to resist certain diseases, one of the most important practical applications of the HGP will be improved medical treatments for those illnesses, as well as for genetic disorders, which are caused by mutations, or changes, in the DNA sequence of specific genes.

Anyone wishing to view the information from the Human Genome Project can find it at http://www.ornl.gov/sci/techresources/Human_Genome/home.shtml.

Chronology

Timeline of Discovery

1834	Joseph Lister discovers shape of red blood cells
1836	General Santa Anna attacks the Alamo
1846	The planet Neptune is discovered
1853	The first potato chips are made
1859	Darwin publishes *The Origin of Species*
1860	The Pony Express begins delivering mail
1861	The Civil War starts
1865	Gregory Mendel publishes his finding on inherited traits of pea plants; slavery is abolished
1879	Walther Fleming discovers chromosomes
1885	W.S. Burroughs invents the first workable adding calculator
1899	The magnetic tape recorder is developed
1900	Mendel's experiments are "rediscovered"
1905	The word "genetics" is coined by William Bateson
1944	DNA is identified as containing genetic information
1955	Watson and Crick reveal the structure of DNA
1977	Fred Sanger develops DNA sequencing technology
1986	Polymerase chain reaction, which allows DNA to be duplicated, is developed by Kary Mullis.
1996	Dolly the sheep becomes the first cloned mammal
2001	The human genome sequence is released
2005	DNA testing is becoming a rountine practice on prison inmates to prove their innocence. After 26 years in prison, Luis Diaz was freed in August by DNA testing.

Chapter Notes

Chapter 1   A New Kind of Fingerprint

1. Sue Corrigan, *The Mail on Sunday*, "Watching the detectives; Police are using ever more sophisticated methods to track down criminals. And, as a new BBC series reveals, Britain leads the world," 7/15/2001.
http://www.highbeam.com/library/doc3.asp?DOCID=1G1:76583044&num=8&ctrlInfo=Round9c%3AProd%3ASR%3AResult&ao=

2. Ibid.

3. Barry Scheck. Congressional Testimony
http://www.highbeam.com/library/doc3.asp?DOCID=1P1:53745027&num=5&ctrlInfo=Round9c%3AProd%3ASR%3AResult&ao=

Chapter 2   The Mystery of Heredity

1. Dr. Martin Hewlett, "From Mendel to Biotechnology: A Critical Look at the Historical Development and Philosophical Foundations of Modern Biology."
http://www.mcb.arizona.edu/Hewlett/mjhpaper.html

Chapter 3   A Fateful Decision

1. John Marr, *Girders in the Sand* (Suffolk, UK: Exile, 2000).
http://www.2from.com/exile/miescher.htm

Chapter 4   Nuclein

1. Ernst Mayr, *The Growth of Biological Thought: Diversity, Evolution, and Inheritance* (Cambridge, MA: Belknap Press, 1982), p. 808.

2. Horace F. Judson, *The Eighth Day of Creation: Makers of the Revolution in Biology*, 1996.
http://www.fmi.ch/downloads/about_fmi/Who%20was%20Friedrich%20Miescher.pdf

Chapter 5   Miescher's Legacy

1. Rosie Mestel, "Breakthrough on DNA is 50 Years Old," *The Cincinnati Post*, 2/28/2003.

2. Professor P. Reichard, Nobel Presentation Speech
http://nobelprize.org/medicine/laureates/1968/press.html

Glossary

Acid	(Aah-sid): One of many kinds of chemical compounds with a sour taste that can react with a chemical base to form a salt.
Adenine	(Aah-duh-neen): One of the four bases that make up DNA. Adenine always pairs with thymine.
Amino Acid	(uh-MEEN-oh Aah-sid): The smallest unit of a protein.
Canton	(CAN-tun): A region in Switzerland similar to a state.
Chromosome	(CHROME-uh-sowm): The part of the cell located in the nucleus that contains DNA. Human beings have 23 pairs of chromosomes.
Cytoplasm	(SIGH-to-plazem): The clear, jelly-like substance that fills cells.
Cytosine	(SITE-uh-zeen): One of the four bases that make up DNA. Cytosine always pairs with guanine.
DNA/deoxyribonucleic acid	(dee-OX-ee-RYE-bow-new-CLAY-ick Aah-sid): A double stranded molecule that contains the genetic code and that is made up of two base pairs: adenine-thymine and cytosine-guanine, or AT and CG.
Enzyme	(EN-zym): A protein that can increase a specific chemical reaction without being changed or consumed in the process.
Gene	(jeen): The unit of heredity. A sequence of nucleotides.
Genetics	(juh-NEH-ticks): The study of inherited traits.
Heredity	(hair-EH-di-tee): The passing of traits in both plants and animals via the genes.
Histology	(hiss-TAH-luh-gee): Study of the microscopic structure of tissue.
Lipid	(LIH-pid): Another word for fat.
Nucleotide	(NEW-clee-oh-tide): A molecule made from three smaller molecules: A base, a sugar, and a phosphate.
Nucleus	(NEW-clee-us): A sphere shaped mass within a cell that is enclosed by a membrane.
Protein	(PRO-teen): A large molecule made up of one or more chains of amino acids.
Ribosome	(RYE-bow-sowm): The part of the cell where proteins are made.
RNA/ribonucleic acid	(RYE-bow-new-CLAY-ick Aah-sid): A long chain of nucleotides involved in making proteins. RNA "carries" information about the DNA.
Species	(SPEE-sees): A distinct class of living creature with distinguishing features.

46

For Further Reading

For Young Adults

Fridell, Ron. *DNA Fingerprinting: The Ultimate Identity*. London: Franklin Watts, 2001.

Silverstein, Alvin. *DNA, Niigata City, Japan*. 21 Century Books, 2002.

Wiilcox, Frank. J. DNA, *The Thread of Life*. Minneapolis: Lerner Publishing Group, 1988.

Works Consulted

Bada, Jeffrey and Christopher Wills. *The Spark of Life: Darwin and the Primeval Soul*. Cambridge, MA: Perseus Books, 2000.

Hughes, Arthur. *A History of Cytology*. London: Abelard-Schuman, 1959.

Judson, Horace F. *The Eighth Day of Creation: Makers of the Revolution in Biology*. New York: Touchstone Books, 1980.

Lagerkvist, Ulf. *DNA Pioneers and Their Legacy*. New Haven, CT: Yale University Press, 1998.

Marr, John Marr, *Girders in the Sand*. Suffolk, UK: Exile, 2000.

Mayr, Ernst. *The Growth of Biological Thought: Diversity, Evolution, and Inheritance*. Cambridge, MA: Belknap Press, 1982.

Moore, Ruth. *The Coil of Life: The Story of the Great Discoveries in the Life Sciences*. New York: Knopf, 1961.

Corrigan, Sue. *The Mail on Sunday*, "Watching the detectives; Police are using ever more sophisticated methods to track down criminals. And, as a new BBC series reveals, Britain leads the world," 7/15/2001

http://www.highbeam.com/library/doc3.asp?DOCID=1G1:76583044&num=8&ctrInfo=Round9c%AProd%3ASR%3AResult&ao=

Hewlett, Dr. Martin, "From Mendel to Biotechnology: A Critical Look at the Historical Development and Philosophical Foundations of Modern Biology."

http://www.mcb.arizona.edu/Hewlett/mjhpaper.html

Mestel, Rosie, "Breakthrough on DNA is 50 Years Old," *The Cincinnati Post*, 2/28/2003.

Miescher Friedrich. "Ueber die chemische Zusammensetzung der Eiterzellen." Hoppe-Seyler's medicinisch-chemische Untersuchungen 4 (1871)

http://www.fmi.ch/members/marilyn.vaccaro/ewww/dna.pioneer.excerpt.htm

Miescher Friedrich. "Die Spermatozoen einiger Wirbeltiere." Verhandlungen der Naturforschenden Gesellschaft in Basel 6 (1874)

http://www.fmi.ch/members/marilyn.vaccaro/ewww/dna.pioneer.excerpt.htm

National DNA Databank—The Birth of DNA Evidence

http://www.nddb-bndg.org/cases/collin_e.htm

Reichard, Professor P. Nobel Presentation Speech

http://nobelprize.org/medicine/laureates/1968/press.html

Scheck, Barry. Congressional Testimony

http://www.highbeam.com/library/doc3.asp?DOCID=1P1:53745027&num=5&ctrInfo=Round9c%3AProd%3AResult&ao=

On the Internet

Who Named It?

http://www.whonamedit.com/doctor.cfm/1754.html

Friedrich Miescher Institute

http://www.fmi.ch/

http://www.fmi.ch/members marilyn.vaccaro/ewww/dna.pioneer.excerpt.htm

http://www.fmi.ch/downloads/about_fmi/Who%20was%20Friedrich%20Miescher.pdf

Index